PAUL DAVIS

FACES

PAUL DAVIS

FACES

with an introduction by Kurt Vonnegut

John and Joseph Kennedy, 1940. Courtesy World Book, Inc.
Copyright © 1966 by Field Enterprises Educational
Corporation, Publishers.

Joe Namath, Courtesy *Playboy.* Copyright © 1971 by *Playboy.*
Daniel Ellsberg. Courtesy *Playboy.* Copyright © 1972 by *Playboy.*
Leadbelly. Courtesy CBS Records, Inc.
Thelonius Monk. Courtesy CBS Records, Inc.
Johnny Mathis. Courtesy CBS Records, Inc.

Manufactured in Japan.

Design: Ian Summers

Published by Friendly Press, Inc., 401 Park Avenue South,
New York City, NY 10016, United States of America.

Library of Congress Cataloging in Publication Data

Davis, Paul, 1938–
Faces.

1. Davis, Paul, 1938–   . 2. Celebrities–Portraits.
I. Title.

ND1329.D39A4   1985   759.13   85-70861

ISBN 0-914919-04-0

In looking back at these pictures, I realize how fortunate I have been to work with some very exceptional art directors, editors and clients, among them Henry Wolf when he was at *Show* magazine, Kenneth Deardoff at the *Evergreen Review*, Allen Hurlburt at *Look*, Irwin Glusker at *Life*, Arthur Paul at *Playboy*, Richard Gangel at *Sports Illustrated*, Milton Glaser and Walter Bernard at *New York* magazine, Cyril Nelson at Dutton Books, Jerry Smokler and John Berg at Columbia Records, Reinhold Schwenk at Case & McGrath Advertising, and Sandra Ruch at Mobil, who gave me the splendid opportunities and real pleasures of working with them, and whose assignments constitute a large part of the work on these pages. Joseph Papp, producer and director of the New York Shakespeare Festival, deserves special acknowledgment for his enduring enthusiasm and support. I am grateful also to Emanuel Azenberg, Corporate Annual Reports, Inc., John Davis, Stanley and Glenda Davis, Anne J. Drellich, Mr. and Mrs. J. Keller, Don Laviano, James and Kate McMullan, Sylvia Mushkin, the R. J. Miller Center for Advertising Art, Playboy, Inc., Gail Sheehy, Wesley Strick, and the World Book, Inc., for generously lending their paintings to be photographed for the book, and to Chieko and Kenji Nishimura of the Nishimura Gallery in Tokyo, and Sandro Manzo of Galleria II Gabbiano, for their extra help. It has been a rewarding experience to work with publishers Stu Waldman and Marty Goldstein of Friendly Press, whose idea this book was, art director Ian Summers, who so handsomely designed it, Peggy Flaum, Liv Reed, Sharon Kaplan, Meggin Chinkel Siefert, Maggie Groening, Celia Brayley, and other members of their staff who kept everything moving smoothly along. Finally, boundless appreciation and gratitude to Myrna, my wife, to whom the book is dedicated, and without whose liberal efforts it could not have materialized. Thank you, all.

P. D.

FOR MYRNA

My friend Paul Davis told me one time that he would never have become an artist if he had known when a boy how many pictures the world already possessed. If he had grown up in Florence, Italy, he went on, he would have assumed that there wasn't wall space anywhere for even one more painting. But he spent his childhood instead in art-starved Oklahoma towns with names like Centrahoma, where he was born on February 10, 1938, and Jenks and Caddo and Hartshorne and Antlers.

His father, Howard Davis, was a Methodist minister who made the customary moves from one parish to another, winding up eventually with a large congregation in Tulsa, not much more of an art capital than Antlers, say. But it was there that Paul was flabbergasted by his junior high school art teacher whose opinion it was that he could draw so well that he could probably make his living as an artist, if he liked. Whether any of his ancestors were similarly gifted cannot be known, since they were almost all preachers or farmers or housewives who were much too busy and sensible to waste time on art, except perhaps for his maternal grandfather, Homer Ulysses Brookhart, who published an independent-minded newspaper in Kansas, and his mother Susan, who herself aspired to a career in typesetting before her marriage, and who encouraged Paul in his outlandish pursuits. There is this, too: when his father was in the Air Force during World War II, from the time Paul was four until he was eight, he enclosed crude but amusing drawings of military life with his letters home, which inspired his son to send back pictures of the home front.

How Paul ever got from Will Rogers High School in Tulsa to New York City has much in common with Dorothy's fictional trip from Kansas to the land of Oz. The tornado that suddenly snatched him up on a day like any other on the prairie was an art contest sponsored by *Scholastic* magazine. He entered it much as he might have bought a lottery ticket—not expecting to win anything, and still not taking seriously the possibility of becoming an artist—and won a full scholarship to the School of Visual Arts on East Twenty-third Street in Manhattan, which at the time might as well have been Oz.

So here was this Oklahoma rube, whose family had expected him to become something normal like a farmer or insurance salesman, transported to a city so foreign to him that he would write home that he might as well be in Europe, so little could he understand. Within three years he had stopped going to classes because he had so much work as an illustrator for *Playboy* and the *Ladies' Home Journal* and the *Saturday Evening Post* and so on. Moral: New York City, while hell for most people, can be incredibly hospitable to gifted people from simply anywhere, if they are dependable.

Thus began the career of the man whose handsome book this is, and about whom I once wrote: "Whenever I walk around New York City and find myself thinking, 'This really is the art capital of the world,' chances are that I have just passed a poster by Paul Davis." I was speaking of those images stuck up willy-nilly and all over town which celebrated stage productions by Joseph Papp's New York Shakespeare Festival— *Hamlet* and *Threepenny Opera* and *The Cherry Orchard* and *For Colored Girls* . . . and on and on. Those of us who saw the plays found the posters doubly marvelous, first as works of art and second as encapsulations of the souls of the plays. They themselves were theater.

Yes – and Paul's work is so theatrical that the portraits in this book which have nothing to do with movies or plays still make their subjects look like actors, like dynamic actors about to reveal with speeches and actions the cores of their public personalities. The Kennedys, for example, are here playing their reputations to the hilt, as are Truman and Johnson and Kissinger and Guevara – and even the notoriously reticent George Orwell. The portrait of my hero Orwell seems particularly magical, since Paul has enlivened his features not with the charisma of the man, of which Orwell is said to have had very little, but with the wit and importance of his writings.

When I think of so many of the portraits we have of famous people from the distant past, which show us what sorts of eyes and mouths and noses those people had when they were totally at rest, and little more, I realize that the pictures here will tell future generations something usually very hard to understand about ancestors: how alive we were.

I am a writer of invented stories, and I said to the artist Saul Steinberg one time that there were some colleagues with whom I could not conduct much of a conversation. It was as though, I said, they belonged to a different profession. He replied that there were two different sorts of artists – those who responded to life itself, and those who responded to the history of their art so far. That sounded right to me, and I now identify Paul Davis as an artist of the first persuasion. Looking at these faces, I am moved to exclaim over the beauty of his responses to life itself.

Kurt Vonnegut
Manhattan, 1985

I am obsessed with faces. There are billions of people in the world, each with two eyes, a nose, and a mouth, and what always fascinates me is that we can tell each other apart. The number of arrangements of eyes, ears, noses, mouths, foreheads, hairlines, cheeks, chins and other features seems to be infinite. Add to these the varieties of expression and the possibilities are endless indeed. And when I meet someone I haven't seen in twenty, maybe thirty years, and instantly recognize him or her, I realize there is another essential aspect of the person that has nothing to do with the physical – call it a soul – that further defines an individual's uniqueness.

From the earliest drawings on, part of what I was able to achieve was a reasonable likeness. The process interested me and came rather naturally. Even after all these years it remains intriguing. Mostly it is a matter of getting all the details correct. A likeness is really an accumulation of detail. I began with Dick Tracy, Mickey Mouse, Pogo and Popeye, graduated at the age of twelve to an oil painting of my father, and have been learning ever since.

When I was a little boy, someone gave me a book called *Fifty Famous Americans*, with drawings in it of Charles Lindbergh, Henry Ford, Babe Ruth, and so on. It was a book I really loved. There was something thrilling about seeing all those people together. It occurs to me that this book is a kind of sequel, and I'd like to think it may come to mean as much to someone else as that book meant to me.

P. D.

FACES

This was used as a cover painting for *The New Republic*. Truman was an everyday kind of man, always sure of himself in a forthright, down-to-earth way.
It was surprising – even incongruous – that someone so familiar could become President of the United States.

Another familiar type from Texas whose music always elicited in me a feeling of sadness and being far from home. This was an album cover for Columbia Records.

This painting served for the book jacket
of the autobiography of this great songwriter
and fellow Okie. His sympathy for ordinary
people came from hard experience.

This painting was for an album cover for Columbia Records. There was a time in the '50s when it was nearly impossible to even neck with a girl without Johnny Mathis singing in the background.

Four of the most unforgettable and hilarious characters created by Gleason, in an illustration for *Show* magazine.

As the cover for *Vista* magazine, this was a send-up of Norman Rockwell. Both Ike and Rockwell suggested the same '50s stability and respectability. My aunt Alma married John Eisenhauer – a cousin of Ike's.

An illustration from a series I painted for
World Book 1966 Year Book to accompany
a reminiscence written by Rose Kennedy.
This is a very young Jack, painted from
an old black-and-white photograph.

This was an illustration for *Look* magazine. Of all who experienced the president's loss, the saddest always seemed to me to be Johnjohn. My own son John was the same age, and I gave this painting to him.

A portrait that was painted after my visit
to Hyannisport to take photos for the
World Book assignment. She was clearly
an indomitable force.

An unpublished dual portrait of the young
Ted Kennedy in which I tried to capture
two different moods.

This painting was for the cover of the *Evergreen Review*. I felt that Robert Kennedy was destined to become president, and so his death changed history.

A painting for *Fortune* magazine of the
Secretary of Defense during the war
in Vietnam.

A previously unpublished painting. There has not been a president in recent memory who was as much fun to caricature as LBJ. He was an irresistible target because he reacted like a stung bull.

An illustration for *Look* magazine commemorated the 1954 U.S. Supreme Court decision to enforce school integration.

This was the cover for the *Evergreen Review* after the Cuban rebel leader's death. The painting was used as a poster in New York subways to advertise the magazine issue. The posters were mutilated, and the magazine's offices were bombed – presumably by Cuban emigrés.

A painting that was for a poster publicizing
a concert to benefit Cesar Chavez's
United Farm Workers in 1968. A long
boycott of California grapes followed.

This was a cover painting for *New York* magazine. After one too many commissions involving violent and threatening images, my request for something more positive brought forth this assignment to portray one of our great mythic heroes – the Coop.

This unpublished painting was for a *Show* magazine story that reported the reclusive actress being glimpsed here and there on the streets of Manhattan. A second version was published, but this one is more interesting.

Paul Davis

A cover painting for *TV Guide* of America's idealized Great Depression family in one of those idealized family portraits. The way we never were.

Another kind of family, this was painted for a *Life* magazine story about Leo Castelli, the influential New York dealer in contemporary art, with real and professional relatives.

This unpublished painting for a record
album cover of the folk singer and political
activist is one of the best portraits
I ever did.

This is a cover portrait for *Rolling Stone*.
Dylan, to me, was one of the most
charismatic and mysterious stars
of the '60s.

The poses in this painting for *Look* magazine were based on movie stills from *Help!* that suggested the Beatles' status as cultural icons who were, at the same time, signalling to us.

An album cover for Columbia Records.
The jazz composer-pianist arrived hours
late for a recording session, after all the
other musicians had given up and gone
home. He winged it and made 'Solo Monk.'

This is a portrait for *Show* magazine of an amazing actor – his face flickers with continually shifting emotions.

This painting was for the cover of a book
of poems by the young Russian
poet who captured the attention of
the American public.

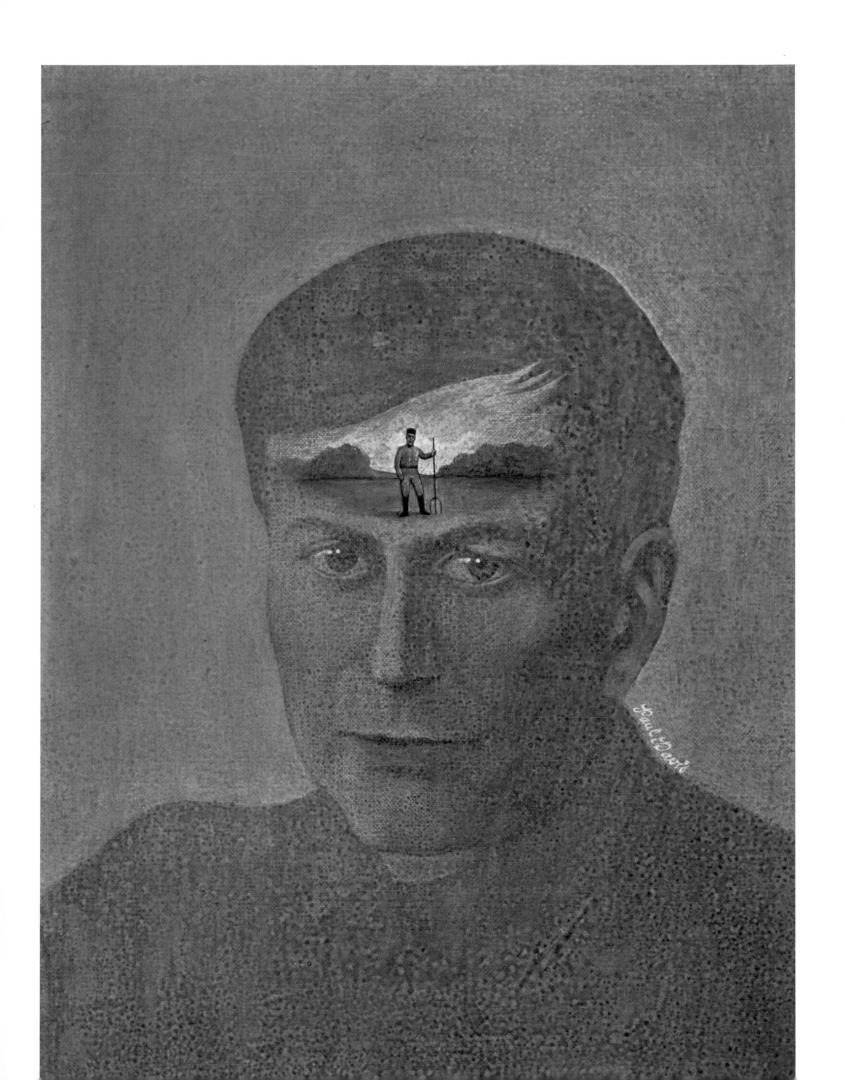

A portrait for *Show* magazine of the
elegant creator of James Bond, painted on
an inelegant piece of wood found in
a barn, while sitting under a tree during
a summer at Woodstock.

This was a painting for *Esquire* of Liz and Dick with assorted children in their happy days. It is somewhat of a lampoon, but an affectionate one.

A portrait for the cover of *New York* magazine depicts 'the last movie tycoon' as the logo for 20th Century Fox.

This unpublished portrait of the consumer
activist was commissioned for the cover
of *Time* magazine. It is painted on a paper
shopping bag.

A portrait for *Look* magazine of the man whose advice Americans raised their children by: Benjamin Spock, M.D. Painted on a baby bottle.

An illustration for *Playboy* of the man who turned over the secret Pentagon papers to *The New York Times* for publication during Nixon's reign.

A portrait for the *Evergreen Review* of the former Harvard professor turned LSD advocate and guru, advising the '60s generation to 'turn on, tune in and drop out.'

This is half of a diptych for *New York* magazine illustrating Gail Sheehy's article, 'Panthermania,' about the militant black group, the Black Panthers.

An unpublished portrait for the cover of *Time*
magazine of the wife of Chairman Mao,
in front of a portrait of her late husband,
after she was overthrown as a member of
the Gang of Four.

This caricature of the wife of the president
of South Vietnam was for a *Ramparts*
magazine story linking the
Diem government with Michigan State
University and the CIA, at the
beginning of America's paranoia
about this strange war.

A painting for the cover of *New York* magazine that was for a story about the part the Mets' victory played in the re-election of Lindsay as mayor of New York City.

This painting, 'Indian Dream,' was commissioned for a portfolio of works by various artists for George McGovern's presidential campaign, in order to illustrate the senator's commitment to American Indian rights.

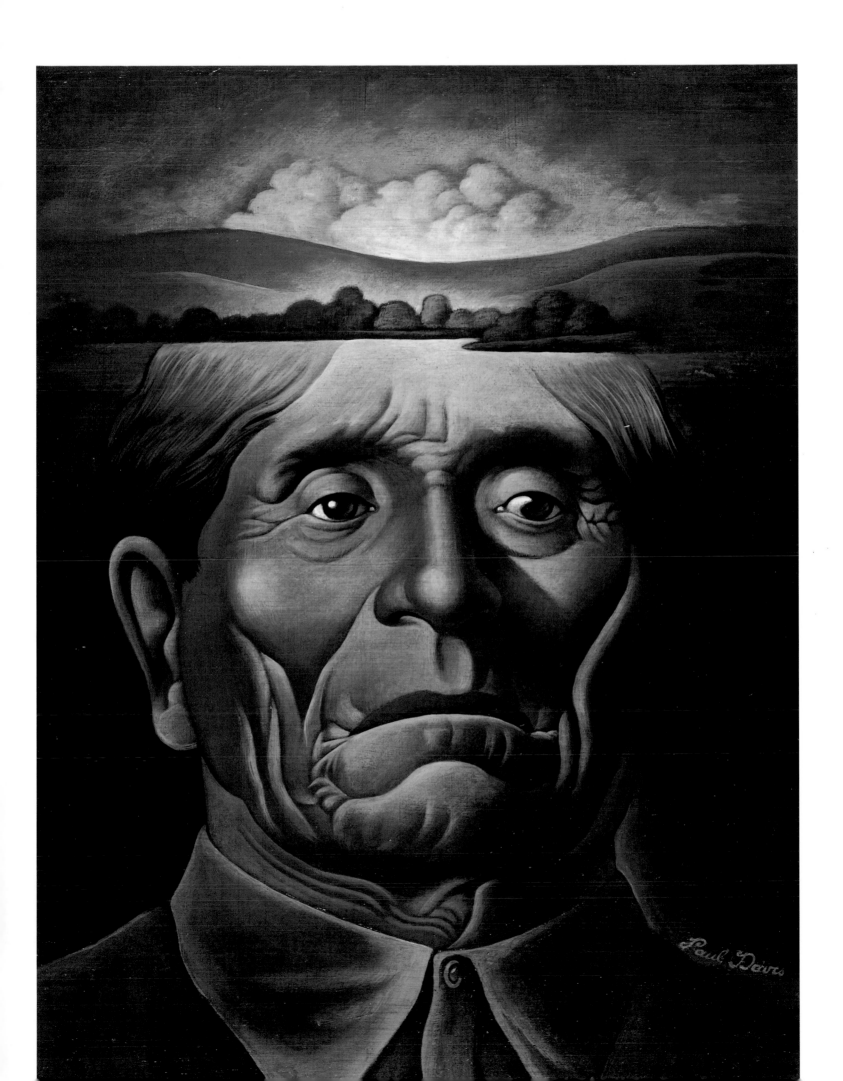

A portrait for the *Evergreen Review*
of the evangelist as spokesman
for President Nixon.

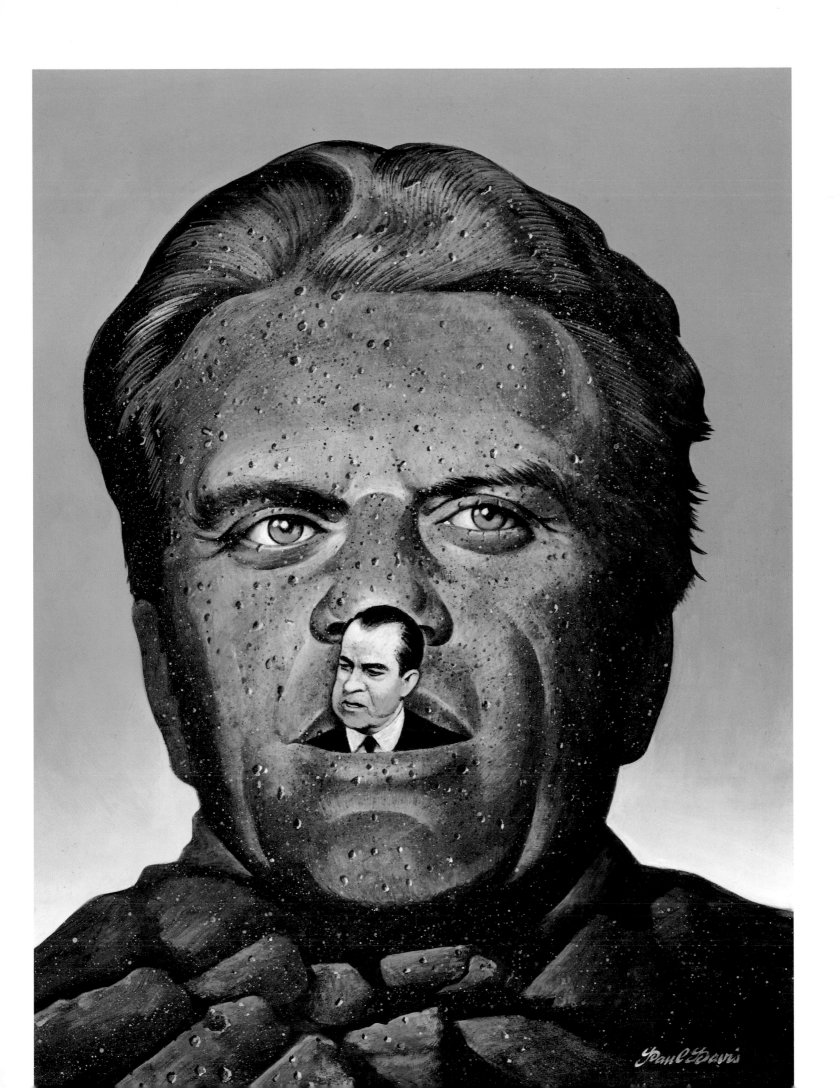

This unpublished cover illustration of Pat and Dick at home was for a *Time* magazine story about government funds being spent for renovations at the couple's California estate.

A painting for the cover of *Ramparts*
magazine depicted Nixon's Secretary
of State as a statue of a Roman senator.

This is an unpublished portrait for the
cover of *Time* magazine. He always struck
me as a decent, unpretentious man.

This painting was commissioned for the cover of *New York* magazine, when the former child movie star became American ambassador to the United Nations.

A portrait of the American Indian activist
and medicine man that was for a film
documentary about his involvement in the
confrontation between Indians and U.S.
government agents at Wounded Knee,
North Dakota.

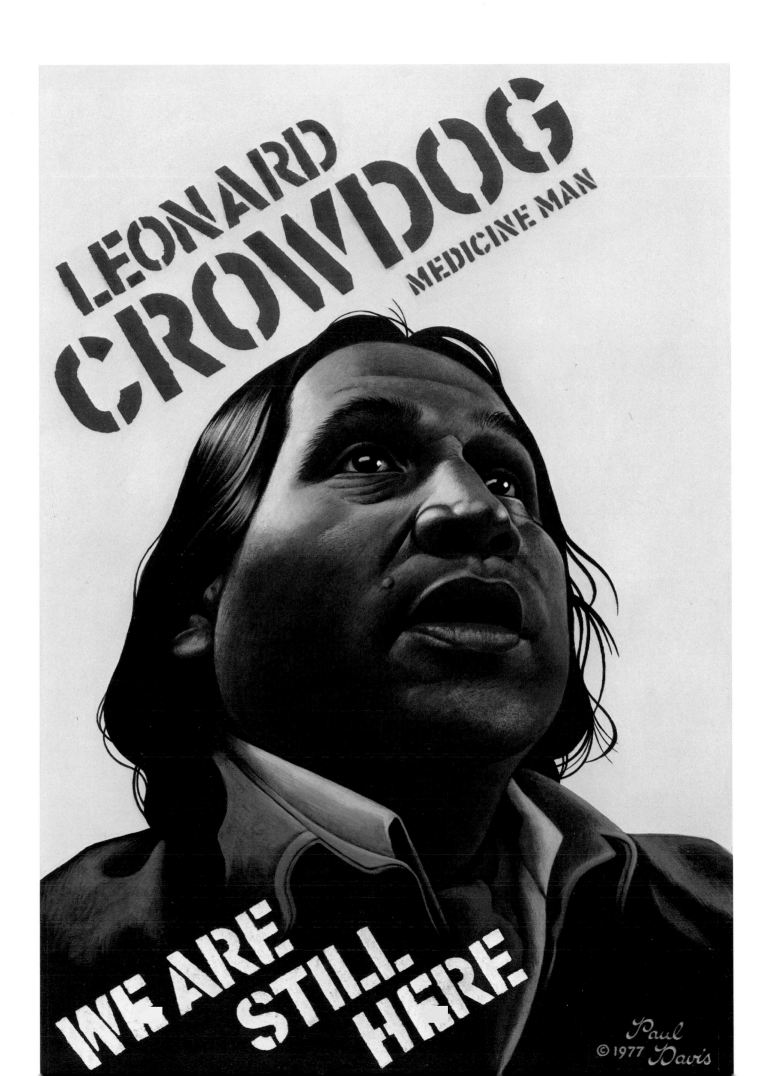

This unpublished portrait of the actor was painted for an advertisement of the movie, *Goin' South*.

This was an illustration for a *Playboy* magazine article about the new career of Broadway Joe as a movie cowboy.

A portrait for a *Sports Illustrated* story on the champion golfer as big business.

This is a recent portrait of one of the outstanding leaders of the women's movement, journalist and founder and editor of *Ms.* magazine.

This unpublished portrait is part of a
series commissioned by the mayor's film
office on moviemaking in New York City.

This cover painting for *Time* magazine
upon publication of his book, *1886,* marked
the first time a third generation had
appeared on its cover: Vidal's father and
grandfather had both appeared in the past.

This is one portrait in a series commissioned for a *Sports Illustrated* story about the prominent horse-racing families of Saratoga, New York.

The portrait of a Russian *refusnik* was for
a poster promoting the annual New York
march in support of Soviet Jews trying to
emigrate from the Soviet Union. This was
painted from a photograph that was given to
me. Accompanying it were the words,
'Take my picture. Show my face. Tell
them I am here.'

This unpublished portrait for *Time* magazine is painted on wood to suggest simple nineteenth-century values untouched by the facts of modern life.

An unpublished painting for *Time*
magazine on the subject of the elderly
in the United States.

Paul Davis

This portrait was for a poster celebrating 'Broadway for Bradley,' a re-election campaign event for the New Jersey senator, former basketball player and Rhodes scholar.

A recent portrait of the former member
of Congress from New York who in 1984
became the first woman candidate for
vice president of the United States.

This portrait was for a surprise poster
celebrating Papp's engagement as a
singer and performer at The Ballroom in
Soho. The producer of the New York
Shakespeare Festival, he is a nurturer of
talent who creates continual opportunities
in theater of astonishing variety and depth.

This portrait and the following eight are part of a series I call actors in their roles. They represent a unique and exciting challenge because I try to capture two personalities in one portrait: the performer and the character he or she is portraying. Olivier as Lear was commissioned for a poster by Mobil for a public television production.

A portrait of author Ntozake Shange
for the New York Shakespeare Festival
Broadway production of her raw, poetic
play, in which she also performed. The
background tiles were meant to look like
subway walls so the lettering would appear
to be real graffiti.

This poster was for the New York
Shakespeare Festival production of
Chekhov's play, directed by André Serban.
The portrait was begun right after a trip
to Japan, and a relationship between the
delicacy and strength of the Japanese prints
I saw and that of Miss Worth immediately
suggested itself.

A poster for a New York Shakespeare
Festival production. Mack is one of those
vivid, enduring characters, nasty and cruel.
Julia, in reality his essential opposite, has
large eyes and a sensual mouth that lent
themselves well to the intended effect.

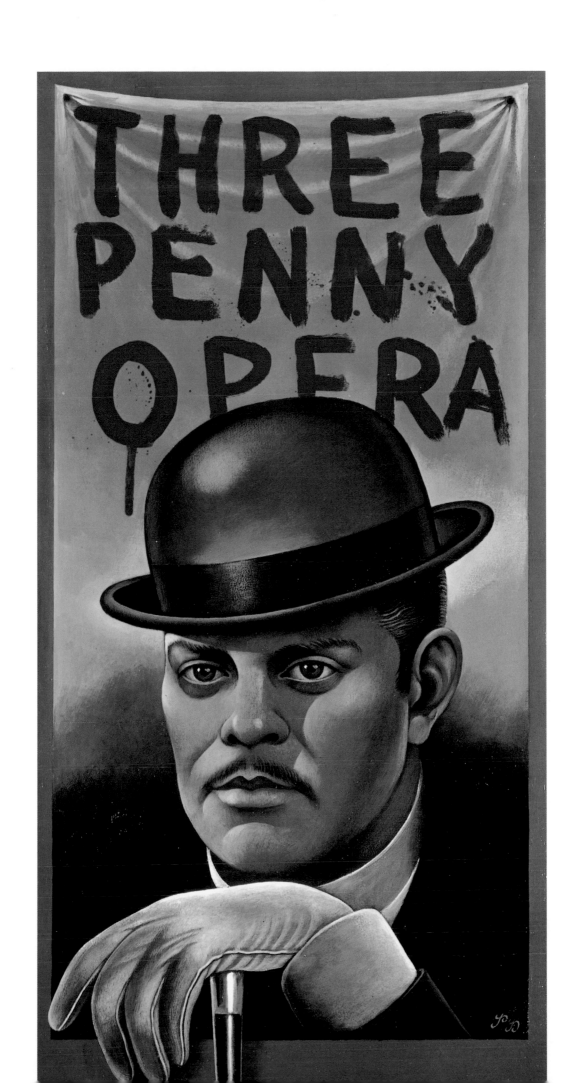

A poster for a New York Shakespeare
Festival production. Instead of an
introspective and passive Hamlet,
Waterston played him in a state
of active madness and rage.

A poster for a New York Shakespeare
Festival production. A really good guy
playing a really bad one again, and Kline's
extraordinary intensity made the portrayal
of evil convincing.

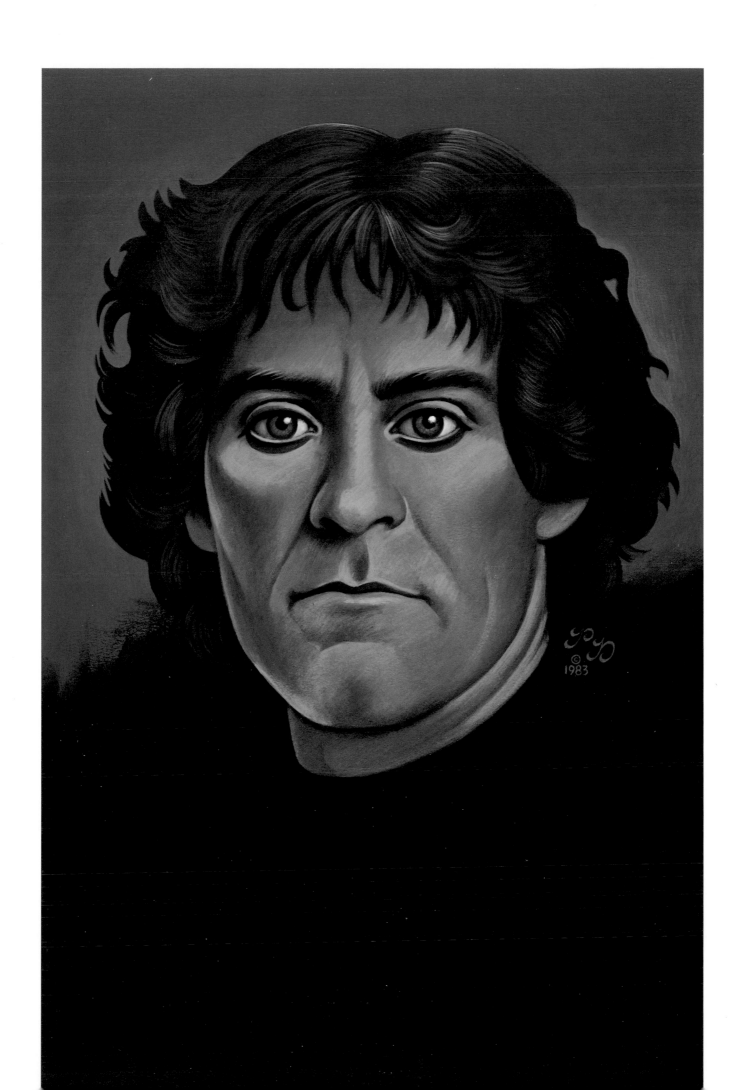

A poster for a New York Shakespeare
Festival production. When Diane Venora
saw the ghost, her face had that expression
of anxiety and tension that a cat has when
it seems to be staring at something no
one can see.

A poster for the New York Shakespeare
Festival production of Elizabeth Swados's
Alice in Concert. Miss Streep is engagingly
self-possessed and straightforward.

I distorted part of the actor's face by
means of a glass to suggest the breakdown
suffered by his character during the course
of E.L. Doctorow's harrowing play.

This poster was for the Boston preview
of a play in which the actor is a father and a
union leader losing his power in both realms.

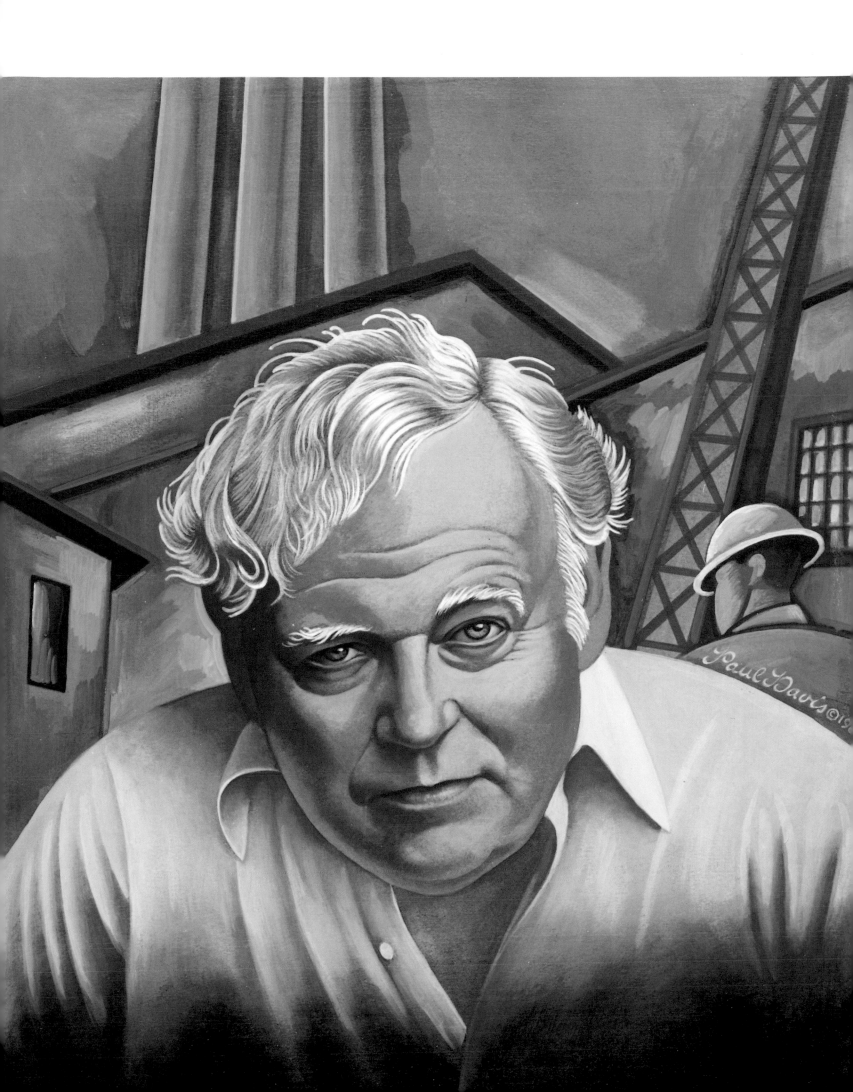

A portrait of the journalist commissioned by *Rolling Stone* magazine in 1976 after his decision to turn over a secret government report to the newspapers caused him to part ways with CBS.

The graphic street dressing of
the '80s, an outgrowth of punk shock,
is celebrated in this portrait of
fashion model and former
fine arts student Pat Kennedy.

© 1985 Paul Davis

A portrait for the cover of *Harper's* magazine to commemorate the author of *1984*. He seemed filled with internal fire, as if about to combust.

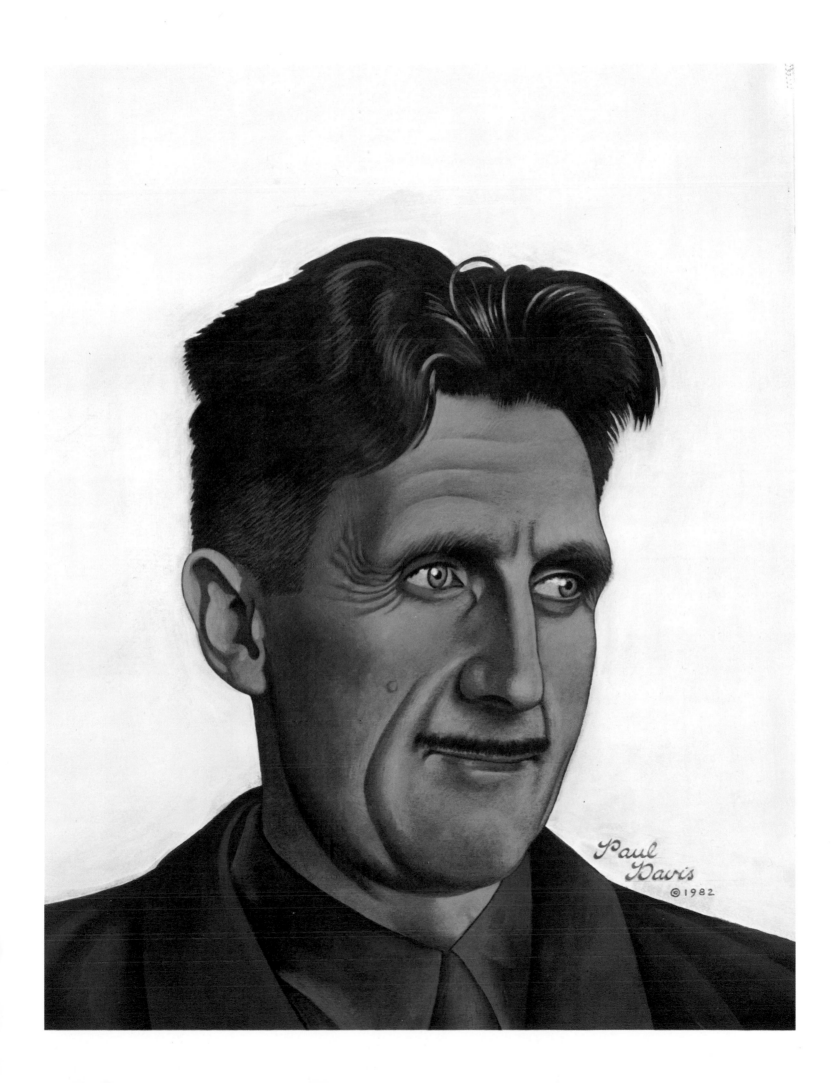